中国农业博物馆 组编

Popular Science Book Series on
A Masterpiece of the Intangible Cultural Heritage of Humanity — The Twenty-Four Solar Terms

Beginning of Spring Sacrificial Rites at Jiuhua Township

Compiled and Edited by China Agricultural Museum

中国农业出版社
China Agriculture Press

北京 Beijing

丛书编委会
Editorial Board of the Book Series

主任 陈　通　刘新录
Directors

副主任 刘北桦　苑　荣
Deputy Directors

委员（按姓氏笔画排列 Ranking by Surname Strokes）
Members
王应德　王晓杰　王彩虹　毛建国　玉　云　皮贵怀
吴玉珍　吴德武　陈　宁　陈红琳　周晓庆　俞茂昊
唐志强　谢小军　巴莫曲布嫫

主编 苑　荣
Editor-in-Chief

副主编 唐志强　韵晓雁　王晓鸣　程晋仓
Deputy Editors-in-Chief

翻译 徐立新
Translator

本书编委会
Editorial Board of This Volume

主任 吴玉珍
Director

副主任 祝亚林
Deputy Director

主编 余仁洪
Editor-in-Chief

副主编 汪筱联　叶裕龙　潘玉光
Deputy Editors-in-Chief

"人类非物质文化遗产代表作——二十四节气"科普丛书
Popular Science Book Series on
A Masterpiece of the Intangible Cultural Heritage of Humanity — The Twenty-Four Solar Terms

序
Foreword

　　中华文化，博大精深，灿若星河，传承有序，绵延不绝。作为人类非物质文化遗产代表、凝结中华文明智慧的"二十四节气"在我国自创立以来，已经传承发展2 000多年。它是中国人观天察地、认知自然所创造发明出的时间知识体系，也是安排农业生产、协调农事活动的基本遵循，更是中国社会顺天应时、指导实践的生活制度。它是中华优秀传统文化中文明成果的典型代表，体现了传统农耕文明的智慧性，彰显了中国人认知宇宙和自然的独特性及其实践活动的丰富性，凸现了中国人与自然和谐相处的哲学思想、文化精神和智慧创造。

"人类非物质文化遗产代表作-----二十四节气"科普丛书
Popular Science Book Series on
A Masterpiece of the Intangible Cultural Heritage of Humanity -- The Twenty-Four Solar Terms

Chinese culture, wide-embracing and profound, brilliant in numerous fields, inherited in an orderly manner, has been developing without interruption. The "Twenty-Four Solar Terms" is a masterpiece of the intangible heritage of humanity and a crystallization of Chinese civilization and wisdom for 2,000 years of its existence. It is a time knowledge system invented by the Chinese people to observe heaven and earth and to learn about nature, the basic principles to organize agricultural production and coordinate farming activities, and also the life system for the Chinese society in its conformation to natural and meteorological timings and in its guidance of daily practices. As the typical representation of the fruit of the best of the traditional Chinese civilization, it embodies the wisdom of traditional farming civilization, reflects the Chinese people's unique interpretation of the universe and nature and the rich practices therein, and highlights their philosophical concepts, cultural spirit and intelligent creativity in their harmonious with nature.

"二十四节气"起源于战国时期，在公元前140年就已经有完整的"二十四节气"记载。从时间上，作为太阳历，早于儒略历（公元前45年）近一个世纪。"二十四节气"较之公历更准确地标识了地球视角的太阳运行规律。农谚就"二十四节气"同公历的关系说道："上半年是六廿一，下半年来八廿三。每月两节日期定，最多不差一两天。"这里所说的"不差"，不是"二十四节气"不准，而是公历有"差"。我们的生活要顺天应时，生活在自然体系之中，就应该把自己看成是包括自然界在内的客观世界的组成部分。无限制地扩大人的能力，破坏自然规律，其后果是难以意料的。

The "Twenty-Four Solar Terms" originated in the Warring States Period. There were already complete records of the "Twenty-Four Solar Terms" in 140 B.C., almost one century earlier than the Gregorian calendar (45 B.C.), also a solar calendar. It is more accurate than the latter in indicating the laws of the sun's movement from the earth's perspective. Agricultural proverbs identify its relationship with the Gregorian calendar: "In the first half of the year, the solar terms fall on the 6th and 21st of each month, and in the second half, they fall on the 8th and 23rd. There are two solar terms in each month, with an adjustment of one or two days." Here, "adjustment" is not the result of inaccuracy of the "Twenty-Four Solar Terms" but errors on the part of the Gregorian calendar. As we need to conform to the natural and meteorological laws and live in natural systems, we should regard ourselves as components of the objective world, including the nature. Expanding human capacity without constraint and disrupting the natural laws may lead to unexpected consequences.

对中国人来说，"二十四节气"是我们时间制度整体的一部分，它是指导我们包括农业在内的创造生活资料的一切活动的时间节律。而我们的情感表达、礼仪等调节人际关系、社会关系的活动则以对月亮运行周期观察为基础的太阴历为节律。我们的传统节日体系大都是以太阴历为依据的。在我们的阴阳合历的整体框架里认识"二十四节气"，领会我们的先辈以置闰的办法精妙恰当地协调二者的对应关系，体现了中华传统文化的精奥和人文精神。

For the Chinese, the "Twenty-Four Solar Terms" is part of our time regime, the time prosody that directs all activities that produce living materials, including agriculture. The time prosody of activities that accommodate interpersonal and social relations, such as our emotional expression and etiquette and protocol, is the lunar calendar based on observations of the moon's periodic movement. Our system of traditional festivals is mostly founded on the lunar calendar. It is advisable to interpret the "Twenty-Four Solar Terms" in the overall framework of the lunisolar calendar, and understand how our ancestors wisely and aptly coordinate the correspondence between the solar and lunar calendars by means of intercalation, which reflects the beauty and humanistic spirit of traditional Chinese culture.

"二十四节气"融合四季，贯穿全年，广为实践，流布全国，影响世界。其作为我国优秀传统文化的典型代表和人类非物质文化遗产代表作项目，富含中国人特有的哲学思想、思维理念和人文精神，具有广泛的参与度和社会影响力，引发世人的关注与探索。2016年11月30日，在文化部非物质文化遗产司指导下，在中国民俗学会支持下，由中国农业博物馆作为牵头单位，联合相关社区单位申报的"二十四节气——中国人通过观察太阳周年运动而形成的时间知识体系及其实践"，被联合国教科文组织列入人类非物质文化遗产代表作名录。这是中国非遗保护工作取得的一项重要成果，也是对外文化交流的一次成功实践。在其带动影响下，全国人民乃至世界人民对"二十四节气"的认知、认同、参与和实践空前提高，进一步彰显和增强了中国人的文化自觉和文化自信。

The "Twenty-Four Solar Terms", integrating the four seasons and covering the whole year, is widely practiced throughout China, with influences on the whole world. As China's best

"人类非物质文化遗产代表作——二十四节气"科普丛书
Popular Science Book Series on
A Masterpiece of the Intangible Cultural Heritage of Humanity -- The Twenty-Four Solar Terms

representative of traditional culture and a masterpiece of the intangible heritage of humanity, it is full of philosophical thoughts, thinking patterns and humanistic spirit unique to the Chinese, enjoying a wide participation and social influence, and commanding attention and exploration from around the world. On Nov. 30, 2016, under the guidance of the Intangible Cultural Heritage Department of the Ministry of Culture and with the support of the China Folklore Society, "the Twenty-Four Solar Terms, knowledge of time and practices developed in China through observation of the sun's annual motion" submitted by the China Agricultural Museum together with related community organizations, was entered onto the list of Masterpieces of the Intangible Cultural Heritage of Humanity by the UNESCO. This was a significant achievement in China's intangible heritage protection, and also a successful practice in cultural exchange. Due to this endeavor and its influence, the people of China and of the world have unprecedentedly heightened their knowledge, identification, participation and practice regarding the "Twenty-Four Solar Terms", which further reflects and enhances Chinese people's cultural awareness and self-confidence.

出版这套"人类非物质文化遗产代表作——二十四节气"科普丛书,有助于在更大更广的范围和层面普及传播节气的相关知识,进一步增强遗产实践社区和群众的自豪感与凝聚力,激发传承保护的自觉性和积极性,扩大关于传统时间知识体系的国际交流与对话,推动人类文明交流互鉴。

The publication of this book series on *A Masterpiece of the Intangible Cultural Heritage of Humanity—The Twenty-Four Solar Terms* shall be conducive to its spread and popularization on a larger scale and in a wider sphere, further enhance the sense of pride and solidarity on the part of the inheritance practice communities and masses, inspire their awareness and initiative in preservation and protection, expand international exchanges and dialogues on traditional time knowledge systems, and promote exchanges and mutual learning between human civilizations.

期望并相信这套丛书能够得到社会各界人士的喜爱。

We sincerely hope and cordially believe that this series will win the hearts of readers of various circles.

谨为序。

Please enjoy your reading of this volume.

刘魁立
Liu Kuili

2019 年 3 月
March 2019

九华立春祭
Beginning of Spring Sacrificial Rites at Jiuhua Township

前 言
Preface

  每年的立春日，作为祈福日，浙江省衢州市柯城区九华乡妙源村外陈自然村，都会在九华梧桐祖殿举行衢州地方传统特有的祭春民俗活动。活动主要有：祭拜春神句芒、迎春接福、赐求五谷丰登、供祭品、扮芒神、焚香迎奉、扎春牛、演戏酬神、踏青、鞭春牛等。鞭春牛是九华立春活动的重要环节，也是整个祭春活动的高潮，表示劝农勤劳和春耕的开始，具有浓郁的衢州地方特色，表达了人们对风调雨顺、五谷丰登、财物丰盛的美好憧憬。

On the day of Beginning of Spring every year, the day of praying for happiness, villagers of the Waichen Natural Village, of the Miaoyuan Village, Jiuhua Township, Kecheng District, Quzhou Municipality, Zhejiang Province, conduct the Spring Sacrifice folk event, a local

九华立春祭
Beginning of Spring Sacrificial Rites at Jiuhua Township

tradition unique in Quzhou, in the Jiuhua Parasol Tree Ancestral Hall. Major rituals include praying to spring god Goumang, ushering in the spring and embracing good luck, praying for blessings of a bumper harvest of all five grains, offering sacrifices, acting as the god of Goumang, burning incense and greeting the spring, handmaking spring bulls, putting on plays to show appreciation to gods, going on spring excursions, and whipping the spring bull. Whipping the spring bull is an important link in the Beginning of Spring rituals in Jiuhua, which brings the whole sacrificial event to the climax. It indicates an encouragement of farming and diligence and signifies the start of spring plowing. It has a rich Quzhou local flavor, and expresses people's high expectations for favorable winds and rains in the year, bumper harvests of all five grains, and bounty finances and resources.

2016年11月30日，联合国教科文组织将"二十四节气"列入人类非物质文化遗产代表作名录，"九华立春祭"作为十个非遗传承保护社区之一，被列入扩展名录。

On November 30, 2016, the UNESCO proclaimed the "Twenty-Four Solar Terms" a Masterpiece of Intangible Cultural Heritage of Humanity, and the "Beginning of Spring Sacrificial Rites at Jiuhua Township" was entered into the extended list of ten intangible heritage preservation communities.

目 录 Contents

序 Foreword

前言 Preface

人文地理 A Cultural and Geographical Overview 1

历史渊源 Historical Origins 9

活动内容 Ritual Events 17

传承谱系 Tree of Inheritance 37

传承保护 Preservation and Protection 40

人文地理
A Cultural and Geographical Overview

　　九华立春祭节气习俗的传承地——浙江省衢州市，位于金（华）衢（州）盆地西部，为钱塘江的发源地。地处浙、闽、赣、皖四省中心，谓"四省通衢"之地，素有"居浙右之上游，控鄱阳之肘腋，制闽越之喉吭，通宣歙之声势"之称。东与省内金华、丽水、杭州三市相交，是浙西有名的生态市，国家级历史文化名城之一。衢州境内河流极大部分属钱塘江水系。衢州地属亚热带季风气候区，全年四季分明，冬夏长、春秋短，光热充足，降水丰沛，气温适中，无霜期长，旱涝明显。

九华立春祭
Beginning of Spring Sacrificial Rites at Jiuhua Township

The home of the folk customs of the Jiuhua Beginning of Spring Sacrificial Rites—Quzhou Municipality, Zhejiang Province, is located in the west of the Jinqu Basin (including Jinhua and Quzhou Municipalities), and here is the source of the Qiantang River. It is situated in the middle of four provinces: Zhejiang, Fujian, Jiangxi and Anhui, a transportation hub of the four. It is thus reputed to "sit on the upper reaches of the Qiantang River in the west of the province, look out on and dominate the Poyang Lake, command the life and death of Fujian and Zhejiang, and echo the popular sentiments of southern Anhui". It neighbors Jinhua, Lishui and Hangzhou Municipalities within Zhejiang Province on the east, and it is a famous ecological municipality in western Zhejiang Province, and a historical and cultural city of national fame. An absolute majority of the rivers within Quzhou belong in the Qiantang River system. The climate here is one of subtropical monsoon climatic zone, with four distinct seasons, the winter and summer long, and spring and autumn short, sunlight and heat plentiful, rainfall abundant, temperature moderate, the frost-free growing season long, and droughts and floods obvious.

古城新貌
Landscape of the City

九华立春祭
Beginning of Spring Sacrificial Rites at Jiuhua Township

庙源溪
Miaoyuan Stream

衢州市柯城区2010年被命名为国家级生态示范区，也是中国"花木之乡""特禽之乡"和"柑橘之乡"。其地貌以山地、丘陵、平原为主，平均海拔只有67米。柯城区生态条件极为优良，森林覆盖率达73.7%，常年空气质量维持在国家一级标准。水资源丰富，市区饮用水源为一级地表水。《水经注》中记载，盛产于这里的柑橘味甜、汁浓、核少、色艳。明代时为朝廷贡品，有"南国佳果"之称。

橘香风貌
Impressive View of Tangerine Trees

九华立春祭
Beginning of Spring Sacrificial Rites at Jiuhua Township

Kecheng District of Quzhou Municipality was awarded the title of National Ecological Model District in 2010, also a town of flowers and trees, one of special fowl species and of tangerines in China. Its landform features mountains, hills, and plains, with an average altitude of 67m. The ecological condition of the Kecheng District is extremely superior. Its forest coverage reaches 73.7%, and it enjoys a national Class I air quality in the whole year. It has abundant water resources. The drinking water of the urban district is Class I surface water. According to *Shui Jing Zhu* (The Commentary on the Water Classic), the tangerines produced in abundance here had a sweet flavor, thick juice, few seeds and bright colors. In the Ming Dynasty, the tangerines were tributes to the royal court, reputed as "best fruit in the south of the country".

衢州有全国仅存的两座孔氏家庙之一——南宗孔氏家庙，于公元1128年，由孔子第四十八代孙孔端友随宋高宗南渡时所建，至今仍有规模盛大的祭孔大典。衢州城南有景色秀丽的烂柯山，素有"道教第八洞天"之称，《晋书》所记载"王质遇仙"的传说即出于此，在《中国名胜词典》《简明不列颠百科全书》中皆有收录。城内现存宋代修筑的古城门六座，还有保存完好程度为江南之最的衢州府城墙。此外，还有《聊斋志异》中所述"衢州三怪"（白布怪、鸭怪、独角怪）出没的县学塘、蛟池塘、古钟楼遗迹等。

Quzhou is home to one of the only two ancestral shrines of the Confucius family in China – the ancestral temple of the southern Confucius clan, which was built in 1128 by Kong Duanyou, the 48th generation of grandson of Confucius when he crossed the Changjiang River with Emperor Gaozong. Quzhou is also home to the Confucius Worshipping Ceremony of a grand scale. It boasts of the Lanke Mountain of breath-taking splendor, located to the south of the city, reputed to be the "Eighth Cave-Heaven of Daoism". The legend of "Wang Zhi's Encounter with Fairies" as

九华立春祭
Beginning of Spring Sacrificial Rites at Jiuhua Township

recorded in *Jin Shu* (The Book of Jin) was set here, and this was an entry in both *A Dictionary of Famous Sights in China* and *Britannica Concise Encyclopedia*. Within the city proper, there are six ancient city gates built in the Song Dynasty, and there are the city walls of the capital of Quzhou Prefecture, which have been the best preserved in southeast China. Besides, there are also the relic sites of County School Pool, Jiaochi Pond and Ancient Bell Tower which are said to be frequented by the "Three Monsters of Quzhou" (White Cloth Monster, Duck Monster, and Single Horn Monster) as narrated in *Liao Zhai Zhi Yi* (Strange Tales from a Chinese Studio).

古城新貌
New Skyline of the City

九华立春祭
Beginning of Spring Sacrificial Rites at Jiuhua Township

烂柯山
Lanke Mountain

孔庙
Confucius Temple

吴越文化、徽派文化以及福建的八闽文化、客家文化等诸多文化因素融合发展，兼容并蓄，形成了极富特色的衢州地方文化。北部吴语、皖南徽语、江西赣语、闽北建州语等交互影响，形成了复杂多样的方言。其中的江山方言堪称汉语活化石，保留了不少古代词汇，具有很高的研究价值。重要的传统节日除了春节、元宵、清明、端午、七夕、中元、中秋、重阳等传统节日外，衢州人还格外重视过立春、立夏、冬至、六月年、麻糍等节日。

Many cultural elements are integrated here in co-development: Wu-Yue Culture, Anhui Culture, and Bamin Culture and Hakka Culture in Fujian Province, and thus have evolved the unique Quzhou local culture. It is also here that the northern Wu dialect, southern Anhui dialect, Gan dialect of Jiangxi, and Jianzhou dialect of northern Fujian interact with each other and give rise to a very complex dialect system here. The Jiangshan dialect here is held in high regard, reputed to be the living fossil of the Chinese language, as many ancient Chinese vocabulary items have been preserved in it with very high research values. Apart from such traditional festivals as Spring Festival, Lantern Festival, Qingming (Tomb-Sweeping) Festival, Dragonboat Festival, Double Seventh Festival, Hungry Ghost Festival, Mid-Autumn Festival, and Double Ninth Festival, Quzhou residents also pay special attention to festivals of Beginning of Spring, Beginning of Summer, Winter Solstice, New Year Celebration in the Sixth Month, and Fried Glutinous Pudding Festival.

历史渊源
Historical Origins

在中国的很多地方，民间都有立春日祭春神的习俗。关于春神的传说很多，有一说认为春神就是句芒神，很多地方都有在立春日祭祀句芒神的习俗。句芒又称勾芒、木正、木帝，是一个鸟身人面、乘两龙的形象。他是中国古代民间神话中的木神，主管树木的发芽生长，辅佐东方上帝，即青帝太皞。中国古代阴阳五行中，东方主木，为青色，主生发，主春。我国民间对句芒神的祭祀远在周朝时就有，那时设有

九华立春祭
Beginning of Spring Sacrificial Rites at Jiuhua Township

东堂，年年都会举行迎春祭礼仪式，这种祈求农业丰收的祭祀习俗一直延续到清末民初。

In many places in China, people have the custom of worshiping the god of spring on the day of Beginning of Spring. There are varied legends about the god of spring. In some stories, he is indeed the god of Goumang, and thus in many places the god of Goumang is worshipped on the day of Beginning of Spring. Also known as Goumang (with a variant spelling in Chinese), Muzheng (god of the palace of the element of wood), and Mudi (god of trees), he has been portrayed as having the body of a bird, the face of a person, and riding two dragons. According to ancient folklore, he is the god of trees, supervising the germination and growth of trees and plants in assistance with the oriental heavenly emperor, i.e. the blue emperor Taihao. In the ancient Chinese system of five elements of *yin* and *yang*, the east is where wood resides and is characterized by blue, in charge of germination and the

春神句芒
Goumang — The God of Spring

spring season. Folk sacrifices to the god of Goumang could be traced back as early as the Zhou Dynasty, when the East Hall was designated and the spring-welcoming worshipping rituals were held every year. The custom of sacrificial offerings to pray for bumper agricultural harvests continued till the end of the Qing Dynasty and the beginning of the Republic of China.

自秦、汉以来，与立春日和二十四节气对应的节气习俗逐渐形成并得以承续，虽然各节气习俗时有断续，内容形式时有变化，但这些节气习俗一直在一定范围内流行和传播。如衢州梧桐祖殿立春日祈福，这一习俗已在金衢盆地西南边缘山区承传数千年之久，目前主要在衢州市柯城区、衢江区、龙游县、江山市、常山县、开化县及遂昌县等区域流行和传播。习俗内容主要包括：立春日鞭耕牛、迎春接福、赐求五谷丰登、扎春牛、扮芒神、供祭品、焚香迎奉、演戏酬神、插花、踏青。

Since the Qin and Han Dynasties, the festivals and customs that corresponded to the day of Beginning of Spring and all twenty-four solar terms gradually came into being and were passed down. While some customs might have been interrupted in history or the events and formats changed, these festive customs have always been observed and spread to a certain extent in certain areas. For example, the festive custom of praying for blessings at the Parasol Tree Ancestral Hall of Quzhou has been passed down and observed for several thousand years in the mountainous areas at the edge of southwestern Jinhua-Quzhou Basin. Currently it is mainly popular in the area covering Kecheng District, Qujiang District, Longyou County, Jiangshan City, Changshan County, Kaihua County, and Suichang County, all part of the Quzhou Municipality. The celebratory rituals mainly include urging the plowing bull with a whip on the day of Beginning of Spring, ushering in the spring and embracing blessings, praying for bumper harvests of all five grains, handcrafting spring bulls, performing in the role of the god of Goumang, presenting sacrificial offerings, burning incense and greeting the god, putting on plays on stage in appreciation to the god, hanging up newly picked flowers on the door, and going on excursions in spring.

九华立春祭
Beginning of Spring Sacrificial Rites at Jiuhua Township

丁酉年立春祭
Beginning of Spring Worship in the Year of Dingyou (2017)

梧桐祖殿
Parasol Tree Ancestral Hall

九华立春祭
Beginning of Spring Sacrificial Rites at Jiuhua Township

迎宾祭春
Welcoming the Guests and Offering Sacrifices to the Spring

关于立春日祈福的节令习俗在当地古代文献中也多有记载。如自清代起，梧桐祖殿立春鞭春牛的记载就出现在外陈村诸姓氏家谱中。民国《衢县志·风俗志》记载：立春前一日官率僚属迎春于东部，出土牛行鞭春礼。当地衙署最高官员也率属迎春。另载：民间犹鼓吹，送春牛图于家者。

These customs were recorded in many local documents in ancient times. Since the Qing Dynasty, entries about whipping the spring bull at the Parasol Tree Ancestral Hall have been appearing in the family records of the various surnames families in the Waichen Village. According to the *Records of Folk Customs in the Annals of*

九华立春祭
Beginning of Spring Sacrificial Rites at Jiuhua Township

Qu County, on the day before Beginning of Spring, the magistrate would head his subordinate officials to greet the spring in the east, presented the earthen bull and whipped it as part of the ritual. The highest ranking official in the government office would also head his own subordinates to greet the spring. It was also mentioned that the local people celebrated by beating drums and blowing trumpets, and by presenting portraits of spring bulls to individual households as a sign of blessing.

春牛迎春
Spring Bull Welcoming the Spring

九华立春祭
Beginning of Spring Sacrificial Rites at Jiuhua Township

随着2016年"二十四节气"被列入人类非物质文化遗产代表作名录，作为扩展名录中的代表性项目，每年立春日在衢州市柯城区的九华乡妙源村梧桐祖殿举行的立春祭也蜚声远近，全国各地的人们纷至沓来，体验立春日祭句芒神、鞭春牛等节庆习俗盛况。

After the "Twenty-Four Solar Terms" entered the list of masterpieces of Intangible Cultural Heritage of Humanity, the Beginning of Spring Sacrificial Rituals held in the Parasol Tree Ancestral Hall in the Miaoyuan Village, Jiuhua Township, Kecheng District of Quzhou Municipality on the day of Beginning of Spring every year have risen to greater fame far and wide. Visitors from all around the country pour in, to look on the festive spectacles of worshipping the god of Goumang and whipping the spring bull.

少男少女接春
Boys and Girls Ushering in the Spring

九华立春祭
Beginning of Spring Sacrificial Rites at Jiuhua Township

诵读祭文
Reading out the Eulogy

巡村祈福
Going around the Village and Praying for Auspiciousness

活动内容
Ritual Events

　　立春节气的名称，是随着斗纲所指的地方并结合当时的自然气候与物候景观命名而来的。立春排序在二十四节气之首，又是春季的第一个节气。其含义是开始进入春天，"阳和起蛰，品物皆春"。衢州梧桐祖殿之立春鞭春牛和相应的祈福节令习俗有：

The name of the solar term of Beginning of Spring was derived from observations of the direction of the bucket arm of the Big Dipper together with the natural climate and phenology of the time. Beginning of Spring is listed at the head of the twenty-four solar terms, also the first of the spring season. It signifies the entrance into spring – since this day, the sun gets warmer and creatures start to awaken, vitality comes back to the earth and signs of spring will be seen everywhere. The solar term customs of whipping the spring bull and others for praying for blessings, on the day of Beginning of Spring, at the Parasol Tree Ancestral Hall, include the following:

（一）春牛图 Spring Bull Sketch

每年的立春日前，人们要将春牛图准备好。旧时，报春人挨家挨户送春牛图上门。春牛图上有春牛和芒神形象，红纸黑线版印，贴于墙壁上，与春联、门笺相辉映，象征一年伊始，春耕将届之意。民国《衢县志·风俗志》载：民间犹鼓吹，送春牛图于家者。

On the day before Beginning of Spring, spring bull sketches shall be prepared, which include images of spring bulls and the god of Goumang. In old times, the spring announcers presented spring bull sketches to the households from door to door. These were printed on red paper with black threads. When put up on the walls, they reinforced the effect of the spring festival couplets and papercuts on the doors, symbolizing the advent of spring plowing at the start of the whole year. According to the *Records of Folk Customs in the Annals of Qu County*, among the masses, drum beating and trumpet blowing were especially featured, and spring bull sketches were delivered to each household.

九华立春祭
Beginning of Spring Sacrificial Rites at Jiuhua Township

村里长者给村民送上春联和春牛图
Elders Presenting Villagers with Spring Festival Couplets and Spring Bull Sketches

（二）备"迎春接福"案桌
Laying the "Ushering in Spring and Embracing Happiness" Table

案桌由三张八仙桌横向组成。迎春祭祀前，将案桌放置在梧桐祖殿大门外正中位，在案桌前披上写有"迎春接福"字样的红纸（锦），案桌中间放饭甑，盛得尖满，甑后置一杯清茶，左右放置香炉、烛台、青菜、梅花、松柏、竹枝，象征洁净、长青和富足。

九华立春祭
Beginning of Spring Sacrificial Rites at Jiuhua Township

Three square tables are set down linearly. Before the worshiping ceremony, the tables are laid out in the due center outside the main gate of the Parasol Tree Ancestral Hall. On the front of the tables is hung red paper (or brocade) with the words "Ushering in Spring and Embracing Happiness". In the middle of the tables are placed dinner pots filled to the top. Behind the pots is placed one cup of clear tea. On the left and right sides of the pot are laid out an incense burner, candleholders, green vegetables, plum blossoms, pine and cypress branches and bamboo shoots—symbolizing cleanliness, everlasting green and abundance.

贴桌案迎春字幅
Putting Up Red Paper with Spring-Welcoming Calligraphic Characters on Tables

九华立春祭
Beginning of Spring Sacrificial Rites at Jiuhua Township

贴春联
Putting Up the Spring Festival Couplets

准备祭品
Preparing Sacrifices

九华立春祭
Beginning of Spring Sacrificial Rites at Jiuhua Township

准备祭品
Preparing Sacrifices

祭品
Sacrifices

（三）选接春少男少女
Selecting Boys and Girls as Spring Ushers

在 8～12 岁少男少女中各遴选 8 名，作为接春少男少女，配以服饰，按习俗头上戴柳条圈，手提油纸灯笼。交春时刻半个时辰前，站立梧桐祖殿正门外两侧，按左四男四女、右四女四男排列，寓以四时八节，吟唱立春相关诗词、歌曲。

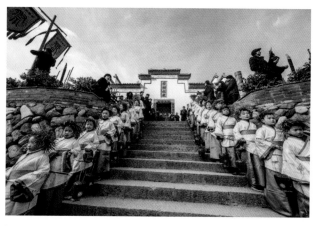

接春少男少女
Boys and Girls as Spring Ushers

Eight young boys and eight young girls at the age of 8-12 are selected as the Spring Ushers, who will wear specially made costumes. Before the rituals of ushering in the spring, they will put on makeup, wear garlands of willow twigs, and carry lanterns made of oil papers. One hour before the exact timing of Beginning of Spring, they stand on both sides of the main gate of the Ancestral Hall with four boys and four girls on each side, representing four seasons and eight representative solar terms of the seasons, recite and sing poems and songs related to Beginning of Spring.

（四）悬挂二十四节气灯笼
Hanging up the Twenty-Four Solar Terms Lanterns

定制二十四节气油纸灯笼，一面上书"迎春接福"，另一面上书二十四节气名称。交春前一天全部悬挂在大殿春神两侧。入夜时点燃蜡烛，交春前再点燃蜡烛。

The twenty-four solar terms oil paper lanterns are made, on one side of which is written "Greeting the Spring and Welcoming Good Fortune" and on the other side of which is written the name of one of the twenty-four solar terms. On the day before Beginning of Spring they are all hung up on both sides of the God of Spring in the Hall. At dusk they are all lit and before the moment of Beginning of Spring they are lit up again.

大殿悬挂着二十四节气油纸灯笼
The "Twenty-Four Solar Term" Oil Paper Lanterns Hung Up in the Hall

（五）鞭耕牛 Whipping the Plowing Bulls

春牛是象征农事的耕牛。立春日，由人扮句芒神，鞭耕牛，由地方官行香之礼，叫作"打春"，以表示劝农勤劳和春耕的开始。

The spring bull is a plowing bull symbolizing farming and agriculture. On the day of Beginning of Spring, someone will play the god of Goumang to whip the plowing bull, and the local magistrate will present incense. The ritual is known as "whipping the spring", indicating an encouragement for farmers and the beginning of spring plowing.

九华立春祭
Beginning of Spring Sacrificial Rites at Jiuhua Township

新春第一耕
First Plowing in Spring

（六）迎春 Welcoming the Spring

在交春时刻前，开主殿门迎春，摆设迎春接福案桌。交春时刻，鸣放鞭炮，梵香行祭礼迎春。

Before the hour of Beginning of Spring, the main gate of the Hall is opened, the tables for bringing in the spring and welcoming good fortune are laid out. At the moment of Beginning of Spring, firecrackers are fired, incense is presented and burned, and rituals are held to greet the spring.

接春
Ushering in the Spring

九华立春祭
Beginning of Spring Sacrificial Rites at Jiuhua Township

开门迎春
Opening the Doors and Ushering in the Spring

迎春祈福
Welcoming the Spring and Praying for Auspiciousness

九华立春祭
Beginning of Spring Sacrificial Rites at Jiuhua Township

（七）抬神迎春赐福
Carrying the God, Greeting the Spring and Receiving Blessings

地方士、农、工、商万民云集，手拈干香，锣鼓班打起《将军令》，细乐班奏起《朝天子》等曲，抬神迎春赐福巡各村。

Local scholars, farmers, craftsmen and business men, all gather together, and hold and roll incense between their palms, when the gong and drum band and the orchestral band start to play the music. The god of Spring is carried on a parade to various villages, to greet the spring and welcome good fortune.

抬神迎春赐福巡游
Parade of Carrying the God of Spring to Welcome the Spring and Pray for Auspiciousness

九华立春祭
Beginning of Spring Sacrificial Rites at Jiuhua Township

（八）踏春、探春 Spring Excursions and Explorations

旧时衢州有立春日踏春之习俗，市民赶到妙源梧桐祖殿，祭拜春神，感受春光，谓之"探春""踏春"。

In old times, people in Quzhou had the customs of going on excursions on the day of Beginning of Spring. Local residents arrive at the Miaoyuan Parasol Tree Ancestral Hall to worship the god of Spring, feel the spring sunlight, and this is known as "spring explorations or excursions".

赏春
Admiring the Spring

九华立春祭
Beginning of Spring Sacrificial Rites at Jiuhua Township

踏春
Spring Excursions

九华立春祭
Beginning of Spring Sacrificial Rites at Jiuhua Township

(九) 采青 Picking Spring Leaves

采集冬青、竹枝、松柏及野菜,谓之"采春"。

Holly, bamboo, pine and cypress leaves, and wild vegetables are picked and tied into small bouquets, which is known as "picking the spring".

采春
Picking Spring Leaves

（十）插春 Hanging up the Spring

将采集的松枝、翠竹各种树枝插在门上，谓之"插春"。

Twigs of pine, bamboo and other trees are put up on the doors and this is known as "hanging up the spring".

（十一）带春 Wearing the Spring

将柳枝、竹枝编成环状带戴在小孩头上，能保四季清健，谓之"带春"。

Willow and bamboo twigs are woven into garlands to be worn on the heads of kids. This can keep them fresh and healthy in four seasons and known as "wearing the spring".

戴春的孩童
Children Wearing the Spring

(十二) 尝青 Tasting the Spring

立春日当天的早餐和午餐必须吃青菜，谓之"尝青"或"咬春"。自古至今，立春日民间还一直流传着咬春习俗。咬春又称吃春盘，人们在立春这一天要吃一些春天的新鲜蔬菜，既为防病，又有迎接新春的意味。寓意吃了以后能聪明伶俐、身强体健、幸福长久。

On the day of Beginning of Spring, green vegetables must be served for breakfast and lunch, and this is known as "tasting the spring" or "biting the spring". Ever since ancient times, there has always been the folk custom of "biting the spring" on the day of Beginning of Spring. "Biting the spring" is also known as "eating a plateful of the spring". On this day, people have been supposed to eat fresh vegetables of the spring, to prevent diseases on the one hand, which symbolizes greeting to the new spring. The implication is that when the vegetables are eaten, people will become clever and intelligent, healthy, strong, and happy ever since.

(十三) 享春福（祝春福）
Enjoying Spring Blessings

立春日晚餐做春饼设夜筵，男女欢溦，谓之享春福。

On the day of Beginning of Spring, spring pancakes are cooked and evening banquets are held. Both men and women enjoy the feast happily together. This is known as "enjoying the spring blessings".

做春饼
Cooking Spring Pancakes

（十四）演戏酬神
Putting on *Xiqu* Performances to Thank Gods

　　每年当立春节来临，都要请戏班子在梧桐祖殿对面戏台上唱三天三夜的戏，给梧桐祖殿和诸位神仙看，名为社戏节。逢年过节如春节、元宵、二月二、清明、端午、中秋、重阳、冬至等，也要演戏，还要邀请亲朋好友前来助兴。

九华立春祭
Beginning of Spring Sacrificial Rites at Jiuhua Township

Every year at the advent of Beginning of Spring, a *Xiqu* team is hired to put on performances on the exquisite stage opposite the Parasol Tree Ancestral Hall for three days and nights, for the viewing of the Parasol Tree Ancestral Hall and the fairies, known as "the temple fair festival". On other major festivals, such as the Spring Festival, the Lantern Festival, Second Day of the Second Month on the Lunar Calendar, the Qingming Festival, the Dragon Boat Festival, the Mid-autumn Festival, the Double Ninth Festival, and Winter Solstice, *Xiqu* performances are also staged. Relatives and friends are invited to join too to enhance the happy ambience.

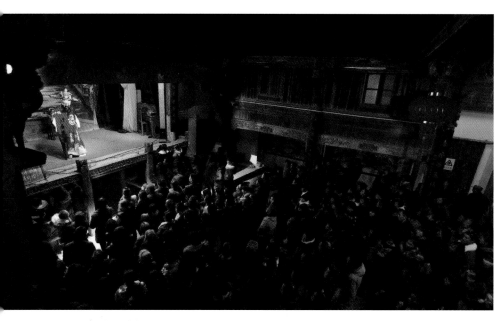

演戏酬神
Performancing Operas to Thank Gods

传承谱系
Tree of Inheritance

　　作为典型的民俗节气节庆文化，衢州九华梧桐祖殿立春祭实际上是在民间以口耳相传的形式一代代地传承。人们从生产、生活和走亲访友的过程中，从传闻中获得"梧桐老佛"的传说故事，并在立春时节进行鞭春牛活动，许多内容已被搜集整理者记录下来。据外陈村《苏氏宗谱》《龚氏宗谱》《傅氏宗谱》等史料记载，大致的传承谱系如下：

As the typical folk custom and festival culture, spring sacrificial rites at the Quzhou Jiuhua Parasol Tree Ancestral Hall have been passed down from generation to generation by word of mouth among the ordinary people. In their agricultural production, daily life and visits to relatives and friends, they learned the legend of the "Old Buddha of Parasol Tree", and whip ped the spring bull at the Beginning of Spring. Most of the rites have been registered in documents. According to historical records of *Family Lineage of the Su's*, *Family Lineage of the Gong's*, and *Family Lineage of the Fu's*, the approximate lineage of the preservation is as follows:

第一代：苏礼本，男，外陈村人，道光二十七年（1847年）房长兼梧桐祖殿执事。

First generation: Su Liben, male, of Waichen Village, the head of the household and the presider of the Parasol Tree Ancestral Hall in the 27th year in the reign of Emperor Daoguang (1847).

第二代：龚元绍，男，外陈村人，光绪十年（1884年）房长兼梧桐祖殿执事。

Second generation: Gong Yuanshao, male, of Waichen Village, head of the household and the presider of the Parasol Tree Ancestral Hall in the 10th year during the reign of Emperor Guangxu (1884).

第三代：苏为授，男，外陈村人，民国八年（1919年）房长兼梧桐祖殿执事。

Third generation: Su Weishou, male, of Waichen Village, head of the household and the presider of the Parasol Tree Ancestral Hall in the 8th year of the Republic of China (1919).

第四代：傅裕农，男，外陈村人，民国三十二年（1943年）房长兼梧桐祖殿执事。

Fourth generation, Fu Yunong, male, of Waichen Village, head of the household and the presider of the Parasol Tree Ancestral Hall in the 32nd year of the Republic of China (1943).

第五代：郑遂安佬，男，1903—1967年，坞口村人，梧桐祖殿执事，师徒传承。

Fifth generation: Zheng Sui'anlao, male, 1903—1967, of Wukou Village, presider at the Parasol Tree Ancestral Hall, inherited the title from his master.

第六代：龚双寿，男，外陈村人，1928年生，1948年为梧桐祖殿的执事，师徒传承。

Sixth generation: Gong Shuangshou, male, of Waichen Village, born in 1928, the presider at the Parasol Tree Ancestral Hall since 1948, inherited the position from his master.

第七代，龚卸龙，男，外陈村人，1965—2008年，父子传承；龚雪龙，男，外陈村人，1968年生，父子传承。

Seventh generation: Gong Yulong, male, of Waichen Village, 1965—2008, inherited the title from his father; Gong Xuelong, male, of Waichen Village, born in 1968, inherited the title from his father.

传承保护
Preservation and Protection

（一）保护内容 Protection Work

1. 做好历史资料的挖掘、整理、研究工作。发掘、收集、记录、整理、归档和出版有关中国立春文化、梧桐祖殿立春祭祀等的资料。建立中国民俗学会立春文化研究中心；协助当地民众传承、规范九华立春祭系列活动；发掘立春文化内涵，出版立春文化系列丛书，打造中国立春文化品牌。

1. Exploration and compilation of the materials and documents has been done. Documents of the Beginning of Spring Culture and the Parasol Tree Ancestral Hall Sacrificial Ceremonies have been explored, uncovered, collected, recorded, edited, classified and published. The Research Center of the Beginning of Spring Culture of the Chinese Folklore Society has been founded; assistance is provided to local people for the preservation and standardization of the Jiuhua Beginning of Spring sacrificial event series; the cultural meaning of the Beginning of Spring culture has been explored and deepened, relative book series have been published, and the Chinese Beginning of Spring cultural brand has been cultivated.

Plaque Unveiling Ceremony for the China Research Center of the Beginning of Spring Culture

2. 完善九华立春祭中扮芒神、迎春、祈福、鞭春、巡村祈福、踏春、探春、采春、尝青、带春、送春牛图等立春民俗活动的相关内容。做好竹编油纸灯笼和春牛图的制作技艺的保护，做好报春人习俗的保护。

2. Work has been done for the standardization and formalization of the Beginning of Spring folk custom events in the Jiuhua Beginning of Spring sacrificial rituals, including such customs as acting the god of Goumang, welcoming the spring, praying for blessings, whipping the spring, making the round of the village for blessings, spring excursions, spring exploration, picking spring leaves, tasting the spring, wearing the spring, and delivering spring bull paintings. Protection has been done regarding craftsmanship of bamboo-woven oil paper lanterns and spring bull paintings, and the spring announcer custom.

3. 修葺、保护好立春祭活动载体——梧桐祖殿。协助做好梧桐祖殿建筑物的保护，完善、恢复西配殿的建筑，修缮建筑本体，整治周边环境。

3. As the site for the Beginning of Spring sacrificial events, the Parasol Tree Ancestral Hall has been repaired and protected. The architecture on the western side hall has been renovated and restored; the Hall itself has been renovated; the surroundings have been straightened and improved.

九华立春祭
Beginning of Spring Sacrificial Rites at Jiuhua Township

梧桐祖殿
Parasol Tree Ancestral Hall

九华立春祭
Beginning of Spring Sacrificial Rites at Jiuhua Township

鞭春牛
Whipping the Spring Bull

诵读喝彩童谣
Reading Aloud Nursery Rhymes

（二）保护措施 Protection Measures

1. 完善九华立春祭保护传承各级组织。成立由衢州市柯城区非物质文化遗产保护工作领导小组成员、梧桐祖殿管委会有关领导和其他相关人员组成的人类非遗二十四节气九华立春祭保护传承工作领导小组，负责日常的保护传承工作。

1. Organizations of different levels for the protection and inheritance of the Jiuhua Beginning of Spring sacrificial rituals have been perfected. The leadership group for the protection and preservation of the Jiuhua Beginning of Spring sacrificial rituals, composed of members of the Intangible Cultural Heritage for Humanity Protection Work Leadership Group of Kecheng District of Quzhou Municipality, leaders of the Administrative Committee of the Parasol Tree Ancestral Hall, and other relevant personnel, directs and coordinates the daily work of protection and inheritance.

2. 成立由九华立春祭的传承人、热心立春祈福民俗的相关人士组成的九华立春祭的民间团体，由梧桐祖殿管委会负责具体工作，保证九华立春祭相关活动的正常持续开展。

2. A civil group for the Jiuhua Beginning of Spring sacrificial rituals has been established, composed of the inheritors and enthusiasts of the rituals. The Administrative Committee of the Parasol Tree Ancestral Hall is responsible for daily operations, so that the ritual events can be practiced as a routine.

3. 成立中国民俗学会立春文化研究中心,通过举办全国立春研讨会、高峰论坛等研究、传承与弘扬中国立春文化与二十四节气文化,扩大立春文化与二十四节气文化的影响与传播,进一步推动民众对立春文化与二十四节气文化的认知与认同,促进文化旅游与文化产业的健康发展。

3. The Research Center of Beginning of Spring Cultural under the Chinese Folklore Society has been established. Through the sponsorship and organization of national Beginning of Spring seminars and summits, the culture of Chinese Beginning of Spring and the twenty-four solar term has been researched, preserved and promoted; perception and identification of the culture on the part of the populace have been promoted; development of cultural travel and cultural industry has been encouraged.

民俗专家指导
Folklore Experts Giving Guidance on the Spot

中国立春文化研究中心成立
The Launch of the China Research Center of Beginning of Spring Culture

4. 成立以九华立春祭管委会为主体的保护传承基金会，多渠道筹措立春文化的保护基金。

4. The Protection and Preservation Foundation has been established composed mainly of the Jiuhua Beginning of Spring Sacrificial Ritual Administrative Committee. Funds for the protection of the Beginning of Spring culture have been raised through different channels.

图书在版编目（CIP）数据

九华立春祭 / 中国农业博物馆组编 . —北京：中国农业出版社，2019.4

（"人类非物质文化遗产代表作——二十四节气"科普丛书）

ISBN 978-7-109-25452-7

Ⅰ.①九… Ⅱ.①中… Ⅲ.①二十四节气－风俗习惯－衢州－通俗读物 Ⅳ.① P462-49 ② K892.18-49

中国版本图书馆 CIP 数据核字（2019）第 079326 号

中国农业出版社出版

（北京市朝阳区麦子店街 18 号楼）

（邮政编码 100125）

责任编辑　张德君　李　晶　司雪飞

北京中科印刷有限公司印刷　新华书店北京发行所发行

2019 年 6 月第 1 版　2019 年 6 月北京第 1 次印刷

开本：787mm×1092mm　1/16　印张：3.75

字数：75 千字

定价：38.00 元

（凡本版图书出现印刷、装订错误，请向出版社发行部调换）